SPACE CHASE

BOOK 1

THE PALMDALE FILES

HAROLD ANDERSON
TOM LYONS

Copyright © 2020 Harold Anderson

All rights reserved. No part of this may be reproduced in any form or by any means without the prior consent of the Author, excepting brief quotes used in reviews.

FIRST EDITION: Published February 8, 2020

Kindle ASIN: B082VWS34V

Paperback ISBN: 9781676976271

THE SECRET TRUTH

This story was published for your entertainment. None of the details contained herein should be considered historically accurate or about real people or events that actually took place… unless you believe.

Any resemblance to actual persons, living or dead, events, or locales is entirely coincidental. The author has disguised and altered numerous facts to protect his identity and the safety of his family.

IN MEMORIAM

This book series is dedicated to all the brave men and women whom I had the fortune of serving with in the United States Air Force.

ACKNOWLEDGEMENTS

Many thanks to my new friend Tom Lyons for your support and encouragement throughout the writing process of this story. And thank you for helping me with all the technical publishing stuff I would've never figured out on my own.

Tremendous thanks to my beautiful granddaughter, "Clarissa." Your graphic and word processing wizardry makes my books look fantastic.

TABLE OF CONTENTS

The Secret Truth iii
In Memoriam v
Acknowledgements vii
Table of Contents ix
Introduction 1
Chapter 1 7
Chapter 2 25
Chapter 3 39
Enjoying This Book? 53
Chapter 4 54
Chapter 5 67
Chapter 6 73
Before You Go 83
Mailing List Sign Up Form 84
About the Authors 85
   Harold Anderson 85
   Tom Lyons 86

x

INTRODUCTION

Life was different in 1936. The world had moved on from World War I but unknowingly spiraled into World War II. Concerned folks helplessly watched in disbelief as brazen leaders like Germany's Hitler, Italy's Mussolini, and Japan's Hirohito rose to power and moved toward a terrible unification—the 1940 Tripartite Pact.

Despite the horrors World War II brought upon the earth, my beloved parents raised me to be an honest, kind man. They wanted me to pursue a joy-filled, meaningful life in pursuit of the American dream, but that dream was taken away when I turned 18 and enlisted with the United States Air Force in 1954—one year before the opening of Area 51.

Though the Second World War had long ended, the cold war between the United States and the Soviet Union reigned supreme and everyone feared World War III would come and go with the swiftness of unilateral nuclear destruction that would end the human race.

After a year of honest service, the Air Force invited me to further my education and specialize in emerging sciences. They subjected me to many physical and psychological examinations that cemented my career in fostering lies and deception—the exact opposite intent of my wholesome, Christian, Minnesota upbringing. My new, secret Air Force role put me at the center of the nation's most classified and dangerous information at Palmdale—an unofficial but critically important organization focused on dealing with "threats from above."

I think it's appropriate to pause here and acknowledge that I cannot and must not disclose, reveal, or hint at my real identity within this content. In fact, I altered the dates to protect my family. Though I am now in my eighties and I have lived a long life, I do not want the truth harming those I love, ever.

Therefore, I will omit certain details that risk revealing any trackable, personal information while fabricating other details. The reasons will be more obvious to some, but my reality understands there are people—just like me—continuously working behind the scenes to suppress certain information from the public eye. I can assure you, I understand the penalty surrounding the revelation of such information to the public.

Thus, I took a simple precaution and labeled this book as fiction. It's crucial that you remember this as you continue to read. With that said, I based everything you are about to read on my recollection of the information about events the

Air Force ordered me to redact or destroy.

I should also note that these events took place over the latter half of the previous century, going back nearly sixty years in my mostly intact memory. Therefore, for the sake of readability, I added minor details here and there, such as timeframes and certain actions e.g. "I was having a coffee when…" and so forth.

I wrote my experiences after speaking with an author acquaintance, Tom Lyons, who writes about his experiences regarding the elusive Bigfoot. After reading his stories—especially *Living Among Bigfoot: First Contact*, in which he mentioned UFOs—I reached out to him and after some encouragement, he agreed to help me write and publish my stories. That said, I agree with Tom when he says, "It couldn't be any more transparent that our government decided long ago that society would function better if certain things were hidden from the public eye. Of course, we witness these cover-ups and diversions all too often, but our busy lifestyles, social interactions, education system, and mainstream media persuade the majority to 'turn the other cheek' and focus instead on the day-to-day tasks that are considered necessary for survival. It's somewhat humorous that a bit of diligent research, combined with some analysis, is all that it takes to recognize that similar agendas have been in effect ever since civilizations first formed."

And likewise, with a slight shift in topic, we agree that "our government consistently denies or disregards the existence of extra-terrestrial life, even though there have been thousands upon thousands of eyewitness accounts. Many theories exist as to why authorities would prefer for these alien beings to remain hidden from the public eye, and I do find a variety of them to be convincing, which suggests a combination of factors could be at play in the suppression of this knowledge."

So, Dear Reader, may my experiences be worthy of your time. Whether or not you believe, I leave you with a quote from a popular television show: the truth is out there.

—Harold

CHAPTER 1

The question of whether we are alone in the universe challenged the very core of my strict Protestant upbringing. For the first twenty years of my life, my unshakable faith denied the existence of evolution. I was a staunch, Bible-based creationist who believed that God Almighty created the universe, the world, and humanity in his image in seven days. (Six, to be exact, because God rested on the seventh day.) Born and raised in a small farming community in southern Minnesota, I entered the world, rather the United States Air Force, as a naïve, inexperienced, and God-fearing young man with no street smarts or understanding about how the world worked.

Dwight D. Eisenhower was President, and he ushered in a grand era of change for the United States; most of which you know about, some of which, you don't. It was my job to make sure certain secrets never made it into your hands, secrets that would terrify, incite chaos or mass rioting, and disrupt the peace and newfound prosperity our great nation grew into over the decades.

During basic training and my subsequent first year of military service, my Air Force superiors quickly discovered I was proficient in mathematics, engineering, and the sciences. I was smart, literate, and I'd like to say well-rounded but my strong faith limited my perception of how the world truly worked. I suspect that's why they chose me, because I was malleable and could learn what they needed me to understand.

Thus, they asked (ordered) me to pursue a

rigorous course of study in those areas for what they claimed was, "the chance of a lifetime." Little did I realize that my studies would grow to include linguistics and various areas of what I call misdirection psychology topics (since there was no official term that I'm aware of) that went far beyond the studies of Freud, Pavlov, and Erikson prevalent in the civilian world at the time. Simply put: I learned how to lie with style and misdirect the entire nation to protect it from itself.

It became clear that the Air Force was grooming our small group—nine down from twenty-five—for some greater purpose, but to what end remained a mystery until the end of our studies. There was no graduation ceremony or any official celebration to mark our completion or readiness for the next step in our careers. Fortunately, the next step came later that day with new orders: pack our belongings from the barracks and prepare to transfer to a secret location the next day by aircraft.

That evening, we laughed and celebrated as much as we could with a few brews at the local pub that had become our unofficial, off-base hangout. Despite my faith's beliefs about the evils of alcohol, my peers had long since convinced me that in moderation, alcohol—including beer—was good for the soul. Besides, with my parents and pastor far away in Minnesota, a few beers couldn't hurt, right?

The locals respected us and for the most part, left us alone. That wasn't always the case with our fellow airmen, who were more than curious about what our dwindling group was doing since they didn't see us out and about on the base, the airfield, the hangars, or with the aircraft. Obviously, we couldn't tell them because we had already been sworn to secrecy and vetted for top secret clearance—something unheard of at our age. The Air Force also warned us not to let this status go to our heads when socializing or working with other airmen. Our superiors expected us to keep our egos in check at all times and remember that humility fostered ultimate greatness.

That night, we didn't imbibe more than one or two beers because our superiors had instilled a key doctrine into our studies: our brains had become critical national assets, and as such, we were special, elite, and chosen for duties everyone else would never comprehend or have the balls to see through. It was a strange message to receive because we knew we weren't being trained for advanced warfare, espionage, or secret missions against the Soviets. But by that night, we knew they had selected us for an entirely new branch of military strategy: the preparation and execution of carefully calculated information manipulation.

The next day, my colleagues and I reported to the airstrip with our belongings and were escorted by Air Force Military Police (MP) to an unmarked hangar. Inside, a Douglas C-124 Globemaster II had been loaded with tarp-covered machinery and we, along with others who seemed equally flummoxed, boarded the aircraft and strapped in. Within the hour we were airborne, and each man prayed we wouldn't be the fourth crash the Globemaster II aircraft series would later suffer that year.

Hours later, we descended and landed at our destination. Though it was hot at Laughlin Air Force Base in Texas, I wasn't prepared for the blistering desert heat of Area 51 in Hiko, Nevada.

Sometime after our arrival and a brief check in, another set of MPs escorted us to the barracks. To our pleasant surprise, these concrete and metal structures were newer and reminiscent of college dormitories. Since there was no basic training associated with this base and—as we soon discovered—the civilian work force was flown in from Las Vegas every day, the accommodations for the permanent on-site Air Force staff were nicer than the group barracks at Laughlin.

"Roger"—obviously not his real name—and I were assigned a unit that was the equivalent of a decent sized two-person college dorm room. After the MP's searched our belongings for contraband—we discovered cameras were forbidden—the MPs left us and told us we had thirty minutes to unpack and then meet them at the main entrance.

With shared levity, we unpacked our belongings (which were limited to what we could carry) and correctly stowed our uniforms and gear, fully expecting a surprise inspection—a regular occurrence at Laughlin despite our supposed elite status. With a few minutes to spare, we explored the dormitory and discovered a large communal shower room at the end of each long hallway. (The building was built like the capital letter H.) Each area had toilet stalls and wall sinks that provided more than enough space for the men to clean up.

When we regrouped at the entrance—located

on the equivalent of the capital letter H's horizontal line—our MP escort transported us by jeep to another building. Our eyes glanced around and took notice of the stark minimalism presented to us. Most of the windowless buildings were solid white or light gray, and none of them were marked. Roger and I thought that was odd, but little did we know we were observing one of the nation's best-kept secrets.

Two hours into our orientation, we learned that we were above classified and entrusted with destroying or falsifying information that could destroy the tense, Cold War-infused atmosphere of Americans terrified of Communism and nuclear war. The last thing they needed to discover was that the real threat to freedom and the American way came from above, from the stars.

That night, my small team of nine didn't speak as we ate dinner together in the mess hall. Around us, other personnel came and went as if nothing unusual ever happened at the base. Later, Roger and I didn't speak as we prepared for bed and climbed into our bunks for the night. The gravity of our situation placed an enormous burden upon our shoulders and minds, as it did with my peers. Clearly, we needed to adjust to our new reality, and fast.

The next morning, our MP escort brought us (again, by jeep) to another unassuming, whitewashed building and ushered us into an amphitheater-style classroom. Upon entry, I was surprised to see Lieutenant General "Smith" (obviously not his real name or rank) and his assistant standing near the podium at the bottom of the amphitheater. The general glanced up from the file folder he was reading as we walked in, and his unknown assistant, a colonel (not his real rank), addressed us: "Gentlemen, please take a seat. We'll begin momentarily."

We silently obeyed (as we were accustomed to following orders) and filed into the second and third rows of seats. The colonel took a seat in the front row and the general finished reading whatever he was reading, allowing the silence to settle upon us. He cleared his throat, closed the file folder, and stepped away from the podium to address us. While I do not recall the exact words Lieutenant General Smith spoke that morning, I remember the shock, confusion, and an overwhelming sense of paranoia that filled my heart and rattled the core of my being. There is no way for me to re-present the general's words to you and hope that you experience the same emotions I did. However, I will attempt to describe what I went through and processed both mentally and emotionally. That said, I will lead with this: Everything he said shattered my Creationist and Bible-based Christian faith to the degree that I no longer remained confident humanity—and humanity alone—was created by God, if he was even real.

"Gentlemen, Good morning. It goes without saying that this meeting is classified above top secret. As far as anyone else is concerned, this meeting never took place and that goes for anything you do while you are here. As you may know, I am Lieutenant General Smith of the United States Air Force and I stand before you today as the representative of a non-existent organization that you are now members of. It is of the utmost importance that you understand this organization does not officially exist and it is never to be mentioned outside of this room or your new workplaces. You do not mention it in your barracks, the pub, the toilet, or when you are convinced you are alone while hiking in the desert. Is that clear, gentlemen?"

We all responded with a unanimous, "yes, Sir."

"The organization you now work for has been codenamed Palmdale. You will begin your work here at Area 51 and remain here until the new Palmdale facility is completed. Of course, no such facility actually exists."

I admit, I was confused because I didn't know if the *non-existent* facility was part of the necessary ruse or if there was, in fact, such a location being prepared for us.

"As you know, roughly 10 years ago in July 1947, a brief media ordeal around a supposed mysterious metal aircraft crashed on a private ranch just northwest of Roswell, New Mexico during a thunderstorm. You may also know that the U.S. Air Force responded and quickly collected the debris for examination. Shortly thereafter, official statements were released debunking the sensational story of a purported alien ship crash."

We all knew what the general spoke of because we studied the psychology of sensationalism, the growing power of the news media, and the strange new threat of conspiracy theories as people distrusted the government more and more. I invite you to remember that in the late 1940s and 1950s, there was no Internet and people's only reliable source of national and global information was distributed through telephone conversations, written letters and other mailed items, the limited number of news networks and television programs, newspapers, and mainstream book and periodical publications. But more interesting and fascinating to the CIA and military, especially during the height of the Cold War, were the smaller, mostly unnoticeable houses of publication that disseminated information the government didn't—and still doesn't—want citizens to have, such as the kind we would soon learn about.

"The Air Force responded to the event near Roswell with the swiftness and professionalism one would expect from this great nation's armed forces. However, the official story we released was a hastily designed fabrication to put the American people at ease. That, gentlemen, is the primary purpose of Palmdale."

Suddenly, everything we had been through started to make sense to me:

- our studies,
- the security clearances,
- the isolation from other standard Air Force operations,
- the ridiculous amount of psychological evaluations,
- the regular patriotism checks (to make sure we were not communists, KGB, or Soviet spies),
- all the seemingly unrelated courses and events we partook in, and
- the lack of clear purpose as to what our superiors expected us to do.

"Your mission has nothing to do with that botched event. Though we will continue to deny the credibility of an alien UFO crash on U.S. soil, we consider the public response and our response to the Roswell event a learning experience and a failure, something we can never recover from and no longer risk. Your mission is to ensure that another media disaster such as Roswell never, ever happens again. Palmdale will become a critical information management team that can intercept, rewrite, and cover up future events so they don't make it into civilian hands. And if they do, we have a contingency plan for that."

The General's assistant went to the door and opened it. He stepped aside, and about 25 clean-cut men in black suits, white shirts, and black ties walked into the amphitheater.

I was flummoxed. How did we not know this team existed? Were they training in parallel with us? I couldn't help but wonder if they were the only other group we would work with in the future as this new Palmdale initiative. As they settled in, we stared at them and they stared at us. They too seemed equally surprised by our existence. The general allowed a few moments of awkward silence to pass as we assessed one another before addressing the large group.

"Gentlemen, Good morning. It goes without saying that this meeting is classified above top secret. As far as anyone else is concerned, this meeting never took place. While it is imperative that soldiers continuously assess their surroundings, allies and foes in the field, I suggest you set aside your egos for a moment and understand that this is one team."

The general repeated the rest of his welcome speech and then picked up with new information. "This unified team, the nonexistent Palmdale organization, represents the brains and the brawn of the nation's best defense: the guardians of its national secrets. When one group is unsuccessful, the other is expected to step in and rectify the situation. This goes both ways, and it is not a numbers game. When intelligence stumbles, muscle steps in. When muscle stumbles, intelligence steps in. As a new endeavor, we will learn and grow along the way. A lack of success is not to be seen as a failure, but an opportunity for growth. For this reason, I expect you to check your egos at the door every day. Leave them in Vegas, leave them at the barracks, leave them wherever you come from and don't bring them to this space. Is that clear?"

"Yes sir," we answered together.

"Excellent. Let it be known that the entire team will not regularly meet like this. We will formalize operations shortly, but given the recent event that needs your expert handling, it's time to see if all the taxpayer money we invested in you was worth it."

Roger and I glanced at one another with contained surprise. There had been an event? Was that the codeword for UFO? Alien crash landing? Something else more bizarre or sinister? Or were we tasked with misdirecting the Soviets while America sought to win the arms race? The Soviets had already launched Sputnik in October 1957, so morale was low because we knew America had lost the space race. When it came to the national defense, America could not accept second place.

And here I thought we were going to do something to assist the nation with the war in Vietnam… I couldn't have been more wrong.

CHAPTER 2

I remember the first time I saw the 1996 movie *Independence Day* in the theaters. I was blown away by the creative imagination of the movie's creators, and at the same time, I couldn't help but wonder if they had inside information from a former Palmdale employee. (It definitely wasn't me.) While the movie focused on a brilliantly executed and fictitious alien species known as the Harvesters, certain details were reminiscent of the truth.

You see, over the years popular comics, newsprint cartoons, and other illustrations often portrayed aliens as emotionless beings with gray skin, oblong heads, and large black eyes. We saw them portrayed as masters of space travel who had figured out so much about space and the galaxy that humanity had yet to discover. However, in reality, they are much like us.

For example, they are not immune to the lethal, cold void of space. Like us, they must wear protective gear—a spacesuit, if you will—to survive in space. And like our astronauts, they wear their spacesuits when piloting their small craft to protect against a sudden loss of pressure or their life-supporting atmosphere.

The fictitious Harvester race received Hollywood's amazing visual effects touch and were vilified with the appearance of evil, angry bugs. These "space locusts" cleverly played off our fears and heightened the average movie viewer's response by inciting a visceral reaction to these gross, monster-like aliens who were hellbent on our destruction so they could harvest the planet's resources. To add to their disgusting appearance, Hollywood gave what turned out to be their spacesuits a number of tentacles that whipped back and forth in a frenzy when they defended themselves. Their disgusting appearance immediately conjured images of the legendary kraken, giant squid, and other scary octopus-like monsters and linked them to the invading aliens, further inciting a time-tested and natural revulsion to the Harvesters and the need to root for humanity. On top of that, they added a viscous slime buffer between the Harvester body and its spacesuit that was revealed to the audience during an autopsy scene. (They thought the alien was dead, but it wasn't.)

Why am I describing this to you? Hopefully, so you can understand the amazing imagination needed to create such a believable, realistic story with a compelling enemy the equally amazing heroes could respond to. (As an Air Force man, I was particularly drawn to the strong, brave character of President Whitmore portrayed by Bill Pullman.)

At Palmdale, I quickly discovered our primary task mimicked that of Hollywood's best creative minds. You see, it was our job to use the same, brilliant, creative strategies, but to do the exact opposite: downplay the fantastic truth as absurd, impossible, and non-existent with equally amazing and believable scripts, but that wasn't always easy.

In the film, the character Russell Casse, played by Randy Quaid, claimed he was abducted in 1986 by aliens who conducted several experiments on him that led to his traumatic belief the Harvesters were planning to destroy humanity. The character eloquently portrayed a stereotypical, crazy version of an alien conspiracy theorist who believed the United States government had been hiding secrets for years. In real life, the government believed those kinds of folks were equally harmless and detrimental to national security and the state of peace America had enjoyed since World War II despite the looming Soviet threat.

"Gentlemen," the general continued, "what I am about to divulge will not leave this base. For those of you shuttling back-and-forth from Vegas, you are not to speak of it once you step foot on the tarmac. None of you will speak of it in the presence of others, and if your privacy cannot be guaranteed for certain while on base, you shall not speak of it. Crystal clear?"

"Yes, Sir," we replied in unison.

"What I am about to describe will be referred to as Event 21 Zeta, or E21Z. The numbering system is arbitrary to prevent the civilian world from identifying or counting actual events. Analysts," (that was my group,) "you will be split into two groups. The local group will create a fiction the Infiltrators can use in town. The National group will create a press-worthy version of said fiction. Though you will be divided, you are expected to work together and spin the sensational story the locals are actively discussing."

We all nodded, wondering what was in store for us. What had happened that created the need for such theatrics? Also, who were these Infiltrators?

"Ground Team, you will initiate full HazMat protocols, establish a concealed perimeter, then catalogue and extract what you find and load it into containers that will ferry everything back here via helicopter and tractor trailer."

Half the black suits nodded.

"Infiltrators, you will make your way into the public spaces and debunk the local chatter with whatever fantastic fake-classified story the analysts come up with."

The other half of the men in black nodded this time. Some of them, I realized, were actors in disguise. The first line of defense was to sell the local public a believable tale that would gently encourage them to stop talking about what really happened — which we still didn't know about yet.

After dismissing the black suits, the general split our group into two teams — alpha and beta. I was instructed to link up with Roger, Dale, John, and Kenneth temporarily. (Obviously, those are not their real names.)

MPs escorted us all to a locked, windowless metal door that opened into a twenty-foot hallway that went nowhere. On the left and right sides of the hallway, about fifteen feet down from the first door, two windowless, locked metal doors opened into rooms that appeared to have been recently built. I smelled fresh paint and there were no marks whatsoever on the alabaster white walls. The door to the right was labeled Alpha, and the door to the left was labeled Beta. Each room held six desks arranged in two rows of three and faced the front wall, which was perpendicular to the hallway. The front wall held a large chalkboard that stretched across the width of the room. We were instructed to choose a desk that would be ours for the duration of our time at Area 51.

I immediately chose the back desk on the left in room Alpha, because it was the furthest away from the door and any commotion that might take place at the chalkboard in the front of the room. Also, I would be able to keep my eye on things, not that I didn't trust these people. After all, I had studied with them for a number of years. Although I didn't know what was going on, I believed in our mission but remained skeptical — of what, I was uncertain. Roger selected the desk next to mine, and the other three men made themselves comfortable at the other desks. We sat down, and Dale immediately rearranged his desk because he was a lefty and it was obvious a right-handed person had arranged the materials on the desk.

Each metal desk came with a brand new box of pencils, a package of erasers, paper clips, a stapler and a box of staples, several thick black markers, and an electric typewriter. The wooden chairs were new, and they swiveled. Each desk also came with a leather seat cushion, and I remember thinking that was a special treat because we were accustomed to hard metal Emeco 1006 Navy chairs or various wooden chairs at our previous location.

General Smith stepped into the hallway space between our two doors and out of habit, we all stood at attention and saluted.

"At ease, gentlemen." We relaxed our stance, and he continued speaking. "Anderson, you'll be doing something a little different."

"Yes, Sir," I replied.

"Each of you demonstrated a specific set of skills that we intend to harness in this first operation and the future. Except for Anderson, you will all be handed documents that must be redacted and blacked out before they are released from this base. The only record of what is in those documents will remain on this base in the underground bunkers. Again, I will remind you, whatever you see, read, or discuss must never ever leave Palmdale. Is that clear?"

"Yes, Sir," we all replied. I was stunned. I had no idea Area 51 had underground spaces to store files, evidence, or work in.

"These documents include testimonies, eyewitness accounts, and photographs recently obtained that point to the veracity of E21Z, information that must never leave the area or catch media attention. That is the information the Infiltration Team must combat with the story you will create. In order to create that story, Anderson will accompany the Ground Team and witness the…objects…firsthand. You will bring a notebook and document everything you see with the incredible level of detail you have demonstrated over the past few years. The reason I'm not sending you all is because I want your creative minds to take the documented stories and photographs and match it with Anderson's detailed notes. Is that clear?"

"Yes, Sir," our two teams hesitantly responded. I raised my hand.

"Anderson? Do you have a question?"

"Yes, Sir. Why am I the only one going to the event site, Sir?"

"At present, we believe limiting analyst visual exposure to E21Z will improve the creative process and allow this team to generate a believable story the locals and any nosy reporters will accept. If this decision inhibits the creative process, we will review and adjust accordingly. Palmdale must evolve to deal with new challenges as they arise. Failure is not an option."

"Thank you, Sir," I replied.

The general motioned for me to step forward, and I did. "These MPs will escort you to the Ground Team. Follow their protocols and you'll be fine. You'll have about two hours on site. When you return, debrief your teammates and create a fantastic story we can use to hide the truth. Oh, and if anyone asks, simply tell them you're the analyst. Do not mention Palmdale by name. You may go now."

"Yes, Sir." I turned to the military police who turned and marched back down the hallway and to the outside until they brought me to a waiting jeep. There, another set of MPs drove me from our building to the tarmac were a number of Ground Team men in black were donning hazmat suits and boarding helicopters.

Support personnel approached me with hazmat gear and helped me put it on. Everyone, including the pilots, wore hazmat suits, but we were instructed not to don helmets or switch on our oxygen supplies. Instead of a notebook, they handed me a portable voice recorder to dictate notes because it would survive the decontamination process, whereas a notebook would not. After a brief tutorial on how to use it and several minutes of patient yet nervous waiting, I boarded one of three Chinook helicopters and shortly thereafter, we were airborne. Based on the position of the sun, I deduced the massive helicopters ferried us northwest to the top of a mountain range I later identified as the Quinn Canyon Range.

As we landed, we were instructed to don our helmets and switch on our oxygen supplies. The Ground Team debarked, and I followed, quickly realizing it wasn't their first time at the site. Several plastic tents or huts (decontaminant-safe command units) had been erected and a significant wall of camouflage netting surrounded a wide area and hid what appeared to be the glint of near-midday sunlight on metallic surfaces. Someone had crashed into the mountain, and the Air Force didn't want anyone to know about it.

But who… the Americans or the Soviets?

SPACE CHASE

CHAPTER 3

I followed the Ground Team through the camouflage wall and gasped. Wreckage carved long gashes into the landscape and scraped up the rocks of the mountaintop. If the craft had cleared the mountain top, it would have sailed into the valley beyond and most likely crashed there, hidden from prying eyes. But up here, the glinting wreckage could be seen for miles on a clear day—which is why the Ground Team early responders (I assumed) had covered most of it with camouflage. I had at least a mile of terrain and wreckage to survey, so I didn't waste any time and switched on the voice recorder.

Since I honestly don't remember what I said or how I said it because it was so many years ago, and because I think that would be rather boring to read through, I'll do my best to describe what I saw in story form. Frankly, that's just more interesting to write and I hope, read.

You're probably wondering if, while on the mountainside, I observed an alien craft, and you'd be within your right to ponder such a thing. At first, I'd didn't know what I was seeing except for large amounts of burnt and scorched metal that had torn a crash-path into the mountain. (If you go looking for it today, the Air Force did a fine job of *restoring* the landscape to erase the extremely visible evidence of the crash path.) I quickly realized (and I'm sure dictated to my trusty voice recorder) that the easiest access to the site, the crash path, and the surrounding terrain began at the aft of the crashed craft. Given my perspective (starting at the tail section and looking down the crash path) it was difficult to ascertain what I was seeing, so I made my way through the twisted metal wreckage.

The metal color was a medium dark gray, though parts of it appeared scorched or charred. On the larger sections, which I assumed were hull or fuselage pieces, the scorches appeared in clusters of three, like the points of an equilateral triangle. One section of wreckage appeared to be a wing with a fuel pod attached to it.

I paused, confused.

The design was earthly, and not extra-terrestrial. Looking back, I saw something I had missed because I didn't walk around the wreckage: the ever-familiar American flag logo present on every U.S. Air Force aircraft. Annoyed with my clumsy mistake, I circled back and examined the wreckage with greater interest and more questions.

- Why didn't they tell us—or me—this was one of our craft?

- Why all the event secrecy when it came to Palmdale agents?
- If we truly were the new Palmdale elite, why not give us all the information we needed to do our jobs up front?
- Didn't General Smith just tell us we were above classified?

I distinctly remember frowning because I realized the Air Force didn't know how to handle these situations and not only were we the supposed trusted elite, we were also the guinea pigs.

"Excuse me," I shouted to one of the Ground Team personnel through my hazmat suit. "What am I looking at?"

"And you are?" he asked, eyeing me with suspicion. That perturbed me because I was wearing a hazmat suit and it's not like I could have snuck onto the mountain top, stolen gear and a suit, and walked into the site.

"I'm the analyst," I replied, just as the general had instructed.

The Ground Team member (I didn't know if they carried rank at that point) nodded and said, "It's a Lockheed U-2. One of the new ones. It was shot down a few nights ago."

"Shot down?" I incredulously asked. "By who?" I had no idea what the man was talking about or suggesting because we were in the middle of Nevada, not a war zone.

"You're the analyst, you figure it out."

"Thanks for nothing," I snapped, irritated about the lack of information I had and could receive. I wondered if my team had a clear picture of what happened based on the accounts they were reviewing. Maybe, I thought, they were just as confounded as I was because according to the Ground Team member, this was a very recent event. The only information my team had were the eyewitness accounts of a plane being shot down at night, but that didn't seem right.

I stared at the wreckage and carefully sidestepped to change my perspective. Soon, I recognized the unique U-2 tail section (though in several pieces), confirmed the wing tank configuration I had seen, and soon I saw the mostly intact fuselage and cockpit, though the bottom and nose were badly scraped up. The crash of the U-2 was reminiscent of a botched landing on the mostly flat terrain at the top of the mountain. The port wing was no longer reattached (and I had already passed it), and the starboard wing had buckled. Crashing without landing gear, the craft had slightly rolled to the port side, preventing a thorough examination.

However, I clearly made out the tri-scorch pattern along the hull, and that is what had captivated the Air Force's attention. Some scorch marks seemed to punch holes through the hull, as if armor-piercing rounds had cleanly penetrated the fuselage. Technicians in hazmat suits were studying the scorch marks with Geiger counters and other tools. That meant it probably wasn't the Soviets. How would they have gotten into our air space without detection?

It also wasn't Canada or Mexico, and since no one else could get their aircraft into American airspace without detection, that left one unlikely possibility: Did we shoot down our own plane?

With a maximum cruising altitude of 70,000 feet, the only plane that could catch Lockheed's U-2 was Lockheed's supersonic SR-71 (with a max cruising altitude of 80,000 feet), but that plane didn't exist yet. That meant we didn't shoot it down. So, who or what shot down a U-2 over American soil, and what weapon did they use to puncture the hull?s?

I remember looking up into the clear, blue sky. The quarter moon was rising or falling near the horizon, and I surmised that outside of the bright starlight, it was very dark at night here.

"Looks like he's getting it," the irritating Ground Team member shouted through his hazmat mask, pulling me out of my thoughts.

I turned to him and grimaced, but the sunlight reflecting off our visors prevented us from seeing the other's face clearly. "I don't understand what I'm supposed to record. This is a crashed U.S. aircraft."

The U-2 was still secret back then, and she didn't look like anything else the Air Force was flying, not to mention it could fly higher than anything else and the recently (and laughable) declassified documentation about Area 51 sightings suggests folks regularly mistook the U-2 for a UFO. Even to this day, the very active Area 51 tests experimental aircraft eager eyes want to cite as extraterrestrial.

They're not always wrong.

I approached one of the Geiger counter wielding technicians and asked, "What are these patterns I'm seeing? These scorch marks?"

I was close enough to see the confusion in his eyes, and the young man didn't know if he should answer me or not. "I'm the analyst."

He nodded with understanding. Apparently, they were expecting me and had been told to cooperate. At least something worked in my favor that day.

"We don't know yet."

"Are they radioactive?"

"No, but there's an unusual electromagnetic reading near the larger holes. They don't seem harmful, but what do I know?"

I moved down the fuselage, studying the scorch patterns. There were no signs of metal stress from bullet strikes, which defied explanation. Something had penetrated—or disintegrated—the hull, but I couldn't tell what. Looking back, I discovered something I had missed. Wherever pieces had broken off the plane, the scorch marks and perforation damage was greater, as if weapons fire had been concentrated in those areas to cut the U-2 apart with surgical precision.

- But, what weapon?
- What aircraft?
- Which enemy?
- And what the hell were we supposed to cover up?

"Have you been to the cockpit?" the technician asked.

"No," I replied, shaking my head and spinning around. How could I have been so stupid? I had limited time, and I hadn't been to the most important part of the aircraft. It appeared intact, except for the ventral area near the still-retracted nose gear and possibly the nose cone. The main landing gear, I noticed and dictated, had also not been lowered.

I made my way to the cockpit, casually observing the Ground Team as they placed numbered tags near every piece of wreckage great and small and snapped photos of everything, especially the tri-scorch marks. The nose cone of the U-2 had been crushed during the crash, so the cockpit was lower than it should have been. All I had to do was climb to a higher vantage point to peer inside, which I did.

The cockpit was open, and the pilot's seat was empty. I wondered if the man was brave or stupid for trying to land his bird instead of ejecting and letting it crash into the mountainside. Then again, that might be the story we'd write instead. After dictating what I saw, I approached another Ground Team technician who stood on a ladder that leaned against the starboard side of the cockpit and waved some kind of electronic measuring device around.

"What happened to the pilot?" I asked. When he glanced at me, I added, "I'm the analyst."

"Did they not tell you that?"

"Funny you should say that. No."

"He's missing."

"Seriously?" The casual tone of the technician struck me as adventurous, so I wasn't sure.

"Yeah. He didn't eject and there are no footsteps leading away from the crash site."

"Was the canopy open when you guys got here?"

"Yup, everything is still exactly as we found it."

"Thanks," I said, and the man went back to measuring things. I stepped back and made a mental list of what I had discovered that didn't add up:

- A U-2 was on a secret mission or training flight in Nevada airspace. Or, it was taking off or landing at Area 51.

- Something shot at the U-2 with surgical precision and attempted to cut into pieces, but the unknown weapon seemed only to damage the U-2, not obliterate it. Either the U-2's hull was too strong, or the weapon was low-powered. I believed the latter was the more plausible reality.

- A U-2 aircraft crash-landed on a mountain side. In Nevada. Its pilot did not eject and also did not walk away. Since the canopy structure and the seat straps were intact, I knew he wasn't thrown from the aircraft.

- The fuel tanks didn't appear to rupture and the U-2 didn't explode.

- The Air Force wanted—or needed—to hide A) the crash B) the missing pilot C) the scorch marks D) the impossible attack? E) all the above.

I knew the answer would be E, and I had a

strong hunch I'd be adding a lot more letters by the time I returned to Area 51 and spoke with my colleagues, which it was nearly time to do. New to the process myself, I dictated everything I saw while simultaneously trying not to feel overwhelmed by the mystery. I also made mental notes about organizing my own process. My notes were haphazard and if someone had to transcribe them, the result would be an unprofessional mess.

ENJOYING THIS BOOK?

If you'd like to learn more about my upcoming stories, please sign up for Tom Lyon's mailing list.

Why sign up for Tom's list?

Honestly, I'm just too old to manage a mailing list, and Tom Lyons graciously offered to manage my books with his. I'm grateful for Tom's support and encouragement with the publication of my classified experiences in the Air Force and I enjoy our ongoing partnership.

As an added bonus to my readership, Tom made *Bigfoot Frightening Encounters Vol. 2* FREE to readers who subscribe to his mailing list. To claim your free eBook, head over to http://eepurl.com/dhnspT and click the "FREE BOOK" tab!

And now, enjoy the rest of my story!

CHAPTER 4

As the Chinook carrying the first team and me ascended and flew away from the mountaintop, I took note that the mountain range was behind us, and the crash path pointed away from Area 51. I didn't know if it was significant, but I also wasn't sure if that mattered to me and my job. The helicopter ride back to Area 51 passed in what seemed like minutes. My brain was so busy working through unanswered questions and hypotheses that I hadn't noticed the time fly by. The missing pilot situation flabbergasted me the most.

After we landed, and as expected, the military police escorted me back to the Palmdale workspaces. Although the exterior security door to our structure was locked, the two internal doors between the work rooms were open, and when I stepped into the Alpha room, members of the Beta team jumped up from their desks and filed into our room. Anticipation and extreme curiosity filled the room with tangible energy I could feel pulsating in the air.

"Well? What the hell did you see?" Dale asked.

"A crash site. One of ours. One of the new U-2 aircraft."

"Where?"

"A mountain range a couple hours northwest of here by helicopter. If the pilot had cleared the mountain top, he would have crashed into the valley beyond it. Instead, his plane skidded across the flat rocky top of the mountain."

"That's it?" Jackson (from team beta) asked.

"You didn't see anything else unusual?"

"Well, the plane was in pieces, and there were strange tri-pattern scorch marks at different points along the fuselage."

One of my team members swore, his tone conveying shock and disbelief. "Why?" I asked. "What have you guys been reading? It's really strange that we were separated because I feel like there's so much information we don't have."

A knock at the door surprised us, and we jumped to attention when we saw General Smith and his assistant standing in the open doorway.

"At ease, gentlemen. Mr. Anderson, I trust you took copious notes with your voice recorder?"

"Yes, Sir," I replied.

"Excellent. As this is our first operation together, I recommend we all listen to the playback as a group after lunch."

"Very good, Sir," I answered with a nod.

"See you at 13:00." The general turned, and he and his assistant left us alone.

I looked at Roger. "Well? What's going on?"

"Locals are going crazy with what they're calling a *space chase*. Numerous reports claim two objects were chasing each other in the sky with the pursuit craft firing some kind of blue-green energy at the first craft. Eye witnesses claim the first craft flew with elegance and style, while the pursuit craft haphazardly zipped across the sky, darting left, right, up, and down, hell, all around the first craft like it was desperate to shoot it down."

"Well, that makes sense," I replied.

"What do you mean?"

"A U-2 doesn't have the maneuverability of a fighter craft. Then again, none of our fighters have the ability to dart around the sky like that. It would seem the scorch marks I saw come from those blue-green energy blasts, but that doesn't explain what happened to the pilot."

"What?" several members of my team asked in unison.

"There was no sign of the pilot," I replied.

"Did he eject?

"Nope, the ejection seat is still in the cockpit and the canopy is intact."

"Did he walk or crawl away?"

"I don't think so. They've searched the area and couldn't find him. I'm sure they'll continue to search for him, though."

"Since we're all going to listen to my notes after lunch, I'd really like to read some of the most relevant documents," I said.

Roger nodded. "There's a stack of unscrubbed reports on the left side of my desk you can peruse through. Lotta people saw this thing go down. The sky was clear and visibility was amazing that night."

"How did that many people see this? This took

place at night, right?"

"Someone spotted it at a bar and ran in to grab people. Then, people telephoned one another. That's how the story spread."

"Has it been printed in the local papers yet?"

"Not that we know of, but there isn't much out here. We're a decent jaunt from any major city. Still, the general wants us to provide a statement for the papers and then concoct a story for the Infiltrators to talk about and be accidentally overheard in the bars tonight."

"And do you have some ideas for the story?"

Dale answered. "Yeah, and we've got some thoughts about how to explain it. Now that we know for sure one of ours went down, that adds credibility to the stories should anyone go investigate."

"Great." The two teams split up, and I grabbed some reports from Roger's desk and sat at my own. The reports came from Air Force personnel who lived in the area and rightfully submitted what they themselves saw in the night sky or overheard others discussing.

Almost every report described the blue-green flashes of light behind the U-2, which at the time I didn't realize was key to creating a believable story.

After lunch and two-and-a-half hours later, we finished listening to my dictated notes. Had I known they would have an audience beyond me, I would have done a better job describing the things I saw. You see, I intended to transcribe the notes and add detail that I remembered as I went along. Still, General Smith was happy with my results and reminded us that nothing was to leave our workspaces. As he prepared to leave, I raised my hand.

"You have a question, Anderson?"

"A recommendation, Sir."

"Let's have it."

"Send more of us next time, if time allows. I may have missed things the group could use to create a story."

"I'll take it under advisement."

"Thank you, Sir."

And then the general left. Moments later, we were all working together, brainstorming plausible explanations the locals would believe while acknowledging a small percentage would not believe anything we said. First, we tackled and provided an innocuous explanation for the general to hand over to the papers and local news stations: Last night, the United States Air Force conducted a nighttime training exercise. During the exercise, one of the aircrafts in involved experienced a fuel malfunction. The pilot was unable to return to the airbase, and the malfunction led to the crash landing of his plane.

We kept it brief and used a phraseology that

wouldn't prompt more questions. Amusingly, the most accurate part of the entire statement, "the pilot was unable to return to the airbase," wasn't questioned by anyone because everyone assumed the pilot wasn't in the aircraft when it crashed.

Next, we brainstormed for an hour and finally submitted the first draft of conversational explanation the Infiltrators could talk about that night at the bars. After receiving the general's immediate approval—which was a surprise to us—we were escorted back to the amphitheater and met with the Infiltrators, all of whom were dressed in Air Force fatigues. They'd be noticed in the bars, and the curious patrons would be more likely to listen to their amazing tall-tale simply because they were dressed for the part.

General Smith announced that he called this part of the plan, EDR, or Explain, Destroy, and Recover. To kick things off, he had us explain our fictitious story and provide discussion points, all of which twisted the observation reports and my eyewitness account of the crash damage into one amazing, believable story:

One of the Air Force's new U-2's fuel tanks unexpectedly froze during prolonged high-altitude training mission and cracked. Thus, the plane's improperly mixed and thus highly explosive Jet Propellant Thermally Stable (JPTS) fuel spilled out of the craft. The unusual amount of solar wind activity that night in magnetosphere (that normally causes the Auroras on the northern and southern hemispheres) ignited the fuel, giving it a blue-green appearance as it violently and haphazardly exploded and burned in the sky. Due to the region's distance from the Earth's poles the auroras did not appear, but the igniting JPTS fuel provided the amazing and confusing display of nature's power. Some fuel explosions occurred close to the U-2, damaging its flight control surfaces and ultimately forcing the pilot to crash before returning to the airbase.

After entertaining some preliminary questions,

General Smith instructed the Infiltrator Team to destroy or poke holes in our story. He wanted them to ask questions civilians might ask and rip the story apart. If we could defend it, the fiction was more likely to become truth. And by doing this as a group, everyone was on the same page and could repeat the exact same story, thus creating believability when the locals discussed among themselves or with friends from surroundings towns who saw the mysterious event.

To our delight, the only questions that challenged the story concerned the veracity of solar wind activity that generated the blue-green lights. Without missing a beat, General Smith said he would contact the director of the newly founded National Aeronautics and Space Administration, or NASA, and instructed them to fabricate a confirmation of the Air Force's statement they could issue if necessary.

Then, General Smith informed the analysts that

we would be traveling to the various bars ahead of the Infiltrators so we could observe and report back on the story's acceptance. We were encouraged to participate in the joviality of the bar scene, play darts, dance, and drink, but abstain from getting drunk. The general wanted us to have clear heads so we could inconspicuously analyze and observe. If anyone asked what we did for work, we were to tell them we worked for a general contracting firm that was just hired to build something confidential on the airbase—a ruse that would explain our sudden presence to the locals who wouldn't recognize us.

"And gentlemen," the general said, "I will be there as well using the alias Mark Havisham. I will play the part of the construction foreman, which means I'll be your boss, a reality that should make things easy. However, tonight I am not the general. Tonight, you are not my soldiers. While you will always proudly serve the American flag, the safety and security of the nation is our top priority and we must enact a clever deception to ensure our way of life is not disrupted. So lighten up, loosen up, but don't get ridiculous. Understood?"

"Yes, Sir," we responded in unison.

The session ended, and we were released to our MP escorts, who brought us back to the barracks to change for our night out.

CHAPTER 5

I honestly don't remember the name of the bar I went to that night, and a quick search on the Internet didn't jog my memory, either. I suspect it has been closed for a while, bought and sold several times, torn down, or left to rot in the hot desert sun. However, I do remember the insane feeling of drinking with an Air Force general who, for one night, wasn't a general. Instead, he was a man dressed in blue jeans, a white tank top, and an unbuttoned buffalo plaid flannel shirt.

Let me tell you, the man could throw darts. Although the Infiltrators were busy feigning inebriation and filling the locals' heads with the "real deal" about the previous night's events, some of us enjoyed "going up against the boss" and tried to beat his untouchable high score. When the locals realized we were temporarily stationed at Area 51, they started asking questions about the experiments we saw and tried to corroborate their theories about the space chase. To their disappointment, we had just enough information to confirm—quietly, away from the Infiltrators-playing-soldier—that we knew a plane with a fuel problem had crashed and they were investigating it. I remember wondering if this would lend any credence to the ambitiously fake story the Infiltrators were spinning.

By the end of the night, the story of the fantastic space chase had died, replaced by the juicy details of the fictitious story we had concocted. Although we couldn't hit every bar and location around the Quinn Canyon Range due to travel times and our proximity to Area 51, the insider story proved more interesting than the space chase and within a week, local word-of-mouth did the job for us. Soon, everyone was enraptured with wild ideas and discussing what kind of experimental aircraft secrets the Air Force was testing at Area 51—conversation that still goes on today. General Smith didn't care what fantastical stories they made up on their own, so long as they weren't concerned with the real, unanswered dilemma we had on our hands: Something not-of-this-world shot down one of our aircraft, and we didn't know who, what or why. We were all left with too many uncomfortable questions:

- Assuming extraterrestrial activity, we didn't know if the aggressor acted offensively or defensively. If the former, was it an act of war?

- Could the United States defend itself from an enemy it didn't understand or believe existed?

- Based on the U-2's damaged fuselage, we knew our metals could be destroyed by their weaponry.

- Based on the sanitized and secured eyewitness accounts, the alien space craft were far superior in maneuverability and high-altitude (not to mention deep space) flight as well.

- Was there just one craft, or was there a fleet of them looming just out of sight, either behind the moon or in some kind of space-based aircraft carrier?

- Where was the pilot? Did the aliens kill him or did they have the capability of removing him from the craft? If yes, when did they do it, just before the crash, or high up in the sky?

- If they rescued or saved the pilot, where was he?

- If they kidnapped the pilot, why? To study

us? For what purpose?

The questions kept us up at night for about a week until exhaustion finally demanded we sleep. Even General Smith could see it was bothering us, but he didn't address it. We were supposed to be the best, psychologically and professionally, but the threat from above took its toll on us because we had no way of finding answers until the extraterrestrials made their next move. I personally drew strength from General Smith because I figured that he had to deal with a lot more stress regarding E21Z than we did — although he probably had more time to come to terms with the fact that aliens were real and had visited our planet.

I wish I had more memory and details about the conversations I had with the locals that night and several subsequent nights. We all–with General Smith—decided it was best to keep up the appearance of contracted workers because he was concerned our one-night-only appearance would eventually backfire and rouse suspicion. Honestly, it was great to get away from the top secret work we started reviewing and redacting, most of which concerned experimental data the Air Force wanted to protect from American and Soviet eyes.

CHAPTER 6

Within a week or two, the Ground Team removed all traces of the crashed U-2. They used the cover of night and massive Chinook helicopters to lift and transport the plane's various parts to the nearest accessible road. The fuselage pieces, wings, fuel tanks, and tail section were lowered onto military-grade flatbed trucks and transported down to Area 51, also at night, and stored in an inconspicuous whitewashed hanger for further inspection.

The Ground Team, in conjunction with other soldiers, who must have wondered what the hell they were doing, used jackhammers and sandblasters to deface some of the crash-scarred rocks so no one climbing the mountain or viewing it from a far would see the unnatural scrapes near the summit.

How do I know this?

I am the one who provided the first-pass redaction on the report that described the clean-up process. I assumed the report was going somewhere, but for all I know, the general had us sanitizing the reports prior to storing them below ground so no one in the future could discover the truth of what happened.

In the surrounding towns, the locals (except for an obnoxiously curious few) forgot about the event and life returned to normal. We cut back our visit to those pubs, but maintained a weekly presence on Thursday nights for a few months to drink, laugh, play pool, and throw darts. All the while, we maintained our contractor personas, and the Infiltrators maintained their Air Force personnel appearance and wore their fatigues. Occasionally, General Smith would join us as "Mark Havisham" and our roles became easier to play. But during the day, whether a crisis faced the nation, or we were in a quiet lull, he was all business and didn't goof around. I am marveled at his ability to affect two personalities. Whether it was skill or a shady act of duplicity, it was amazing to behold.

With the impending conclusion of the U-2 incident, or E21Z, I was hard-pressed to understand the continued value of the Palmdale team. After all, how many experimental aircraft on training missions crash or are observed by the general public? There's a reason the Air Force chose to train its pilots and test experimental aircraft in the middle of nowhere. Only a handful of folks, be they isolationists, Native Americans, or those who call the inhospitable desert regions of Nevada, New Mexico home were privileged to witness things no one else could.

You must remember, Dear Reader, that the late 50s and early 60s were a different era than the information age we now live in. Fifty-plus years ago, people relied on word-of-mouth, the telephone, and newspapers and other printed materials to learn about the world around them. Also, though a broad-stroke statement on my part, I believe most folks didn't want to live that far from civilization. However, people today choose to live away from populated areas and enjoy the quiet peace the desert has to offer.

Today, people are more aware of Area 51 and generally accept that the highly classified U. S. Air force installation serves as a training location for our Armed Forces. It is also commonly known that Air Force installations such as White Sands in New Mexico and Nellis and Tonopah in Nevada serve as the home and testing locations for experimental and stealth aircraft such as the F-117 Nighthawk, the F-22 Raptor, and the B-2 Spirit. In fact, while visiting White Sands Desert in New Mexico as a tourist with my nearly adult grandchildren 10 years ago, we observed a fleet of Nighthawks circling overhead and landing at the airbase.

Today, there is much greater interest in alien conspiracy theories and a pervasive notion that the U.S. government is hiding alien technology at Area 51 and some of its other classified locations. (Like I suggested before, did someone leak information that helped create the movie *Independence Day?*)

Information spreads much faster these days through the Internet, cable television, and an insanely driven media machine desperate to provide information as facts before real details are known. I'm talking about the news giants, such as CNN, Fox News, MSNBC, and others... the list goes on. I don't care where you fall on the political spectrum to notice that news stations go bananas over a developing new story and fall head-over-heels trying to hook you with their ability to "fact find" and sell inaccurate information to their viewers before the truth is known. In my humble opinion, CNN is the worst offender. Then again, Fox News anchors will regularly apologize and correct erroneously reported information. Whether you lean left or right, you just can't win with "news" these days. Alas, I digress…

What you probably don't realize exists is the current level of sanitization enforced by the FBI, CIA, Homeland Security, official military organizations, and the unofficial, nonexistent military organizations such as Palmdale. If we can't get there first, you can be darn sure we'll corner the news crews and isolate them if we need to, but you'll never know it as you comfortably watch the news in your living room. And when things get out of control, the Palmdale Ground Team or the Infiltrators, both known by conspiracy theorists as the Men in Black (another wild movie coincidence, or is it?), take care of the situation using whatever means necessary. (Relax, I can tell you with certainty the fantastical alien-human hybrid world of actor Will Smith and the theatrical *Men in Black* movie does not exist.)

Many of the written reports required redaction and sanitization before they could leave the base and travel to other departments, locations, Congress, and even the President's desk. But even then, everything went through General Smith and a group of high-ranking officers from Area 51. Me and my initial team were personally responsible for destroying evidence of a wide variety of offenses, both natural and unnatural.

But it was E21Z, the bizarre space chase and the mysterious disappearance of its pilot, that ignited my crisis of faith. When I couldn't turn the science to answer the questions that plagued my mind, I looked to my Bible prayed. My faith and belief in the Lord had weakened through several years of intense study, training, and not enough time for regular Sunday worship.

Unfortunately, the Lord had no answers for me.

I was stuck.

Everything my faith taught me about the nature of the universe was that God alone created humanity so that we might worship and serve him. We were his special creation, and there was no room for any life beyond our solar system, let alone the planet. E21Z confronted me with the cold hard fact that something extraterrestrial entered our atmosphere and interacted with one of our aircraft.

It would take me years to find a working balance between science and faith, allowing one to inform the other whenever it was necessary and allowing doubt and an appreciation for the impossible to be out of my control. Of course, that would come after years of rejecting God and working through intense anger and feelings of betrayal at the narrow-mindedness and naïve worldview my faith had ultimately created within me. I was unprepared for the world, but I didn't know it until the possibility of the Christian God was ripped out from under me.

Over the years, the Air Force regularly subjected every Palmdale agent to polygraph (lie detector) tests that became increasingly bizarre and complicated as time went on. I remember wondering if we had become guinea pigs for some greater cause such as the verification of whether polygraphs were accurate means of measure. I assumed they were trying to determine if we were Soviet spies, had spoken out of turn, or accidentally shared classified information while socializing with the curious locals. You must admit, being paid to drink at night occasionally and lie to the general public may seem like a lot of fun, but maintaining lifelong secrecy and a sequence of lies *that must be spoken like fact* are honestly two of the most difficult things I have every done next to cavern diving in New

Mexico… which is where the Air Force planned to send me next.

BEFORE YOU GO

Before you go, I'd very much like to say thank you for reading my book. I'm aware you had an endless variety of unexplainable phenomenon books to choose from, but you took a chance on my content. Therefore, please accept my gratitude for reading my book and for sticking with it all the way to the last page.

At this point, I'd like to ask you for a tiny favor. It would mean the world to me if you would leave a review on this book on the Amazon page. Your feedback will help me as I continue to share my experiences that you and others can find entertainment in.

MAILING LIST SIGN UP FORM

Don't forget to sign up for Tom Lyons' *Living Among Bigfoot* email list. Tom and I promise this list will not be used to spam you, but to ensure that you will always receive the first word on any new releases, discounts, or giveaways! All you need to do is simply click the link below and enter your email address.

Visit the link below!

http://eepurl.com/dhnspT

ABOUT THE AUTHORS

Harold Anderson

Harold Anderson is the pen name of an eighty-something government retiree who worked at Palmdale—an unofficial but critically important organization focused on dealing with "threats from above." At Palmdale, Harold was one of a select few responsible for scrubbing, redacting, or destroying top secret records and 'erroneous information' about extra-terrestrial events that made it into the public's hands to preserve national security and ensure those events never received the credible attention they deserved.

With the recent surge in science fiction, the 2019 viral hoax of *Storm Area 51*, the growing interest in national conspiracy theories, the success of websites like *WikiLeaks*, the recently formed United States Space Force, and the looming threat to our planet's safety from above, Harold now chooses to speak out—a decision that will be considered treason (and punishable by death) if his true identity were ever found out.

Tom Lyons

A simple man at heart, **Tom Lyons** lived an ordinary existence for his first 52 years. Native to the great state of Wisconsin, he went through the motions of everyday life, residing near his family and developing a successful online business. The world that he once knew would completely change shortly after moving out west where he was confronted by the allegedly mythical species known as Bigfoot.